Chicago Public Library
R0125845751
Animals of the grasslands.

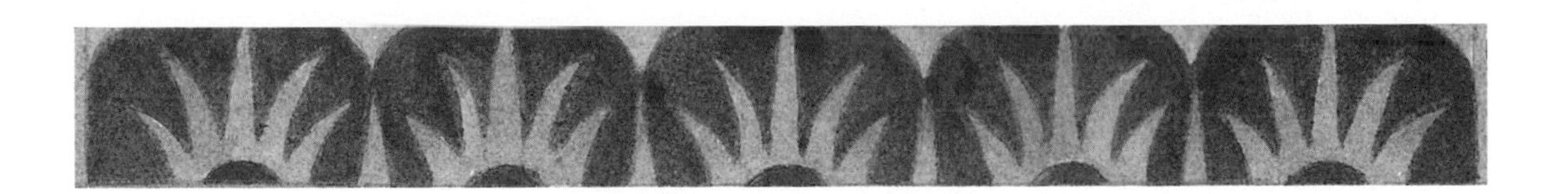

ANIMALS BY HABITAT

ANIMALS OF THE GRASSLANDS

STEPHEN SAVAGE

Titles in the Animals by Habitat series

Animals of the Desert
Animals of the Grasslands
Animals of the Oceans
Animals of the Rain Forest

Published by Raintree Steck-Vaughn Publishers,
an imprint of Steck-Vaughn Company

Library of Congress Cataloging-in-Publication Data
Savage, Stephen.
Animals of the grasslands / Stephen Savage.
p. cm.—(Animals by habitat)
Includes bibliographical references and index.
Summary: Describes the grassland environment and the mammals, birds, reptiles, and invertebrates that live there.
ISBN 0-8172-4752-1
1. Grassland animals—Juvenile literature.
2. Grassland ecology—Juvenile literature.
3. Grasslands—Juvenile literature.
[1. Grassland animals. 2. Grasslands.]
I. Title. II. Series: Savage, Stephen, Animals by Habitat.
QL115.3.S28 1997
591.909'45—dc20 96-27729

Printed in Italy. Bound in the United States.
1 2 3 4 5 6 7 8 9 0 01 00 99 98 97

Habitat Maps
The habitat maps in this series show the general distribution of each animal at a glance.

Picture Acknowledgments

Heather Angel 5 (main picture), 23 (bottom); **Ardea** Clem Haagner 13 (top left), Kenneth W. Fink 15 (top), Jean-Paul Ferrero 20; **Bruce Coleman** Christian Zuber *cover*, M. P. L. Fogden 4, Eric Crichton 9 (bottom left), Peter Davey 9 (top right), Erwin and Penny Bauer 11 (left), Rod Williams 12, Andy Purcell 14, Jane Burton 15 (right), Hermann Brehm 16, Austin James Stevens 17 (top), Jen and Des Bartlett 18, Dr Eckart Pott 19 (bottom left), Harald Lange 19 (top right), Kim Taylor 21 (top), Wayne Lankinen 21 (bottom), John Cancalosi 25 (left), Joe McDowell (top right), Kim Taylor 26, Carol Hughes 27 (left), M. P. L. Fogden 27 (right), George McCarthy 28, Jane Burton 29 (top left); **Frank Lane Picture Agency** Mark Newman 7 (bottom), 10, L. Lee Rue 17 (right), Fritz Pölking 22, M. B. Withers 23 (top), A. A. Riley 24; **Oxford Scientific Films** Steve Turner 5 (inset), Steve Turner 7 (top), 11 (right), Sean Morris 29 (right); **Tony Stone Worldwide** Tim Davis 13 (top right).
All artwork has been supplied by Linden Artists (Clive Spong).

Contents

What Is a Grassland? 4
Grasslands of the World 6
The Grassland Habitat 8
Mammals 10
Birds 20
Reptiles 24
Invertebrates 26

Glossary 30
For Further Reading 31
Notes about Habitats 31
Index 32

Words that are printed in **bold** in the text are explained in the glossary on pages 30–31.

What Is a Grassland?

Natural grassland covers large areas of the Earth and grows in **tropical** and **temperate** climates.

In tropical grasslands, the air is hot and dry for much of the year and the land appears to be scorched by the sun. Trees and bushes are scattered about, providing shade for animals during the hottest part of the day. Watering holes provide much-needed drinking water, but they are also the ideal place for a lion to ambush its **prey**. After the **seasonal** rain, flowers bloom and vegetation grows quickly.

Temperate grasslands contain fewer trees and sometimes none at all, but there are other small plants growing among the grasses. The habitat is flat and exposed to extreme weather conditions. The summer months are usually hot and dry and can reach temperatures of 84°F in some grasslands. In the winter, however, the temperature drops and snow covers much of the ground.

Herds of large **herbivores** roam grasslands, feeding on plants and grasses. Among the grasses, spiders spin their webs, while grasshoppers munch on parts of the grasses and other plants. The battle for food and survival goes on within the grass. At night, the chirruping of noisy insects and the sounds of **predators** hunting their prey can be heard.

Pronghorn antelope graze on the open **prairie** of North America. ▼

Zebras and wildebeests live side by side on the African **savanna**. ▶

Sadly, many grassland animals are now endangered. The African elephant is killed for its ivory tusks, and the rhinoceros is killed for its horns. Big cats are hunted for their beautiful coats, and the American bison is nearly extinct through overhunting. The grasslands themselves are threatened by farming. Domestic sheep, goats, and cattle feed on the same grasses as wild grassland animals. Overgrazing threatens to turn areas of grassland into desert.

▲ The magnificent African elephant is killed by poachers for its ivory tusks.

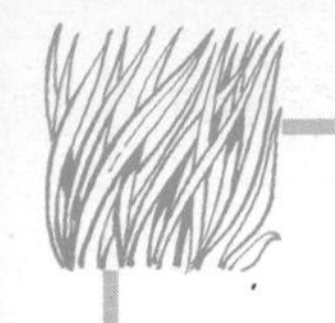

Grasslands of the World

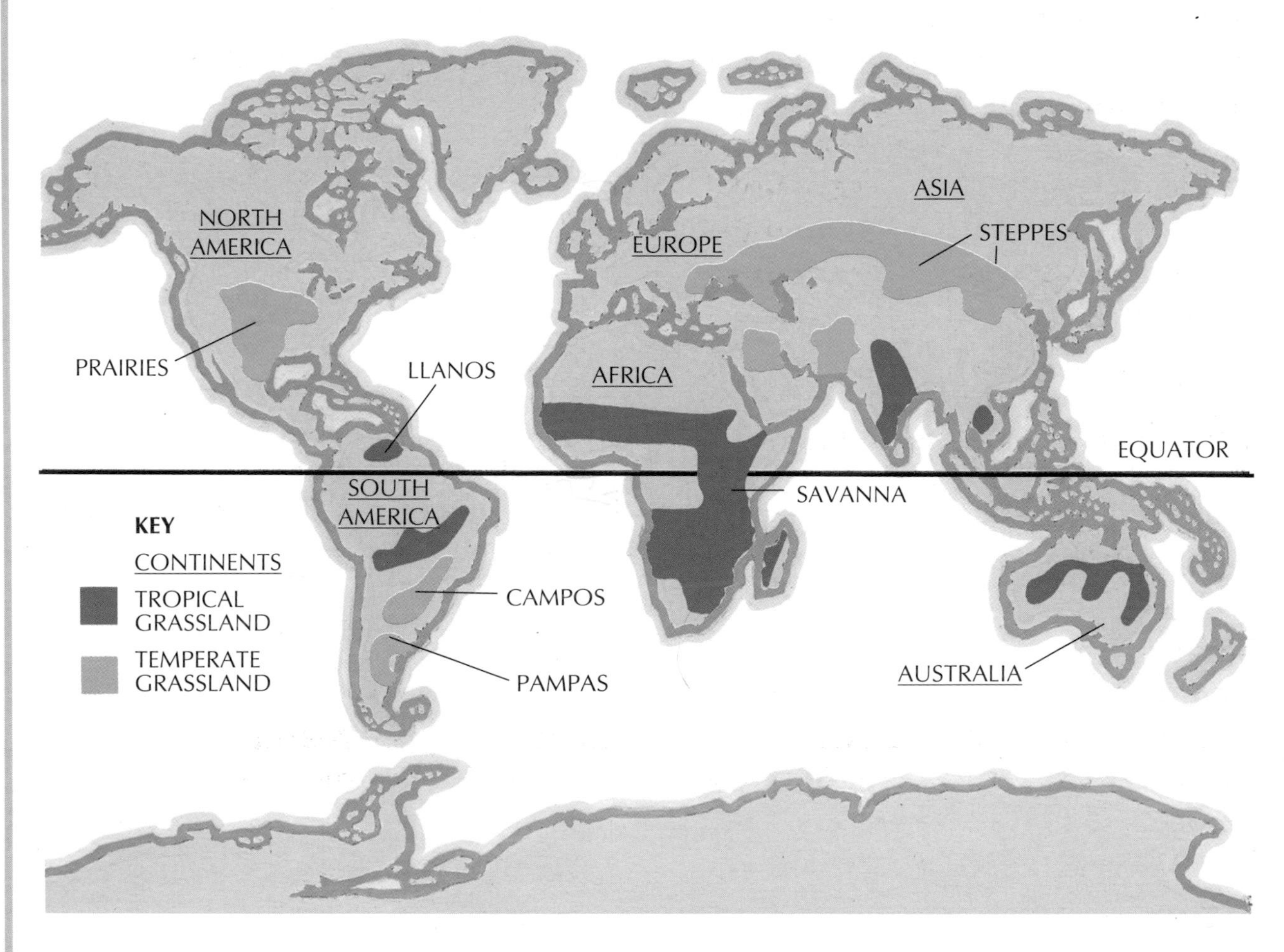

▲ The major natural grasslands of the world

Grassland covers about one-quarter of the Earth's surface. Many grasslands are ancient and have existed for thousands of years. Some grasslands are artificial, created for farming or when woodland has been cleared. When we think of grass, we might think of the lawns around our houses or a school playing field. However, there are about 10,000 different grasses worldwide. Natural grassland is made up of many different types of grasses and can be divided into two main habitats, tropical and temperate.

Tropical grassland grows near the equator and is found in Africa (where it is called savanna), South America, India, southeast Asia, and Australia. The types of grasses that make up these habitats are quite different. During much of the year, the climate is very hot and the grasses have adapted to survive these very dry conditions. In the hot, dry summers, fire may destroy areas of grassland, but this is part of the natural cycle. Although the leaves are burned, the roots are safe beneath the soil, and the grass soon sprouts new leaves.

▲ The wildlife has fled in panic from the fire but will return when the damaged grasses grow back.

Temperate grassland is mainly found in North America, where it is called prairie, and Asia, where it is known as the steppes. Temperate grassland is also found in South America, where it is known as pampas. Temperate grasses are generally shorter than those found in tropical areas. Some parts of the North American prairie are quite wet, while others are dry. The grassland of Asia is a harsh environment to live in. It receives little rain in the summer, and the ground is frozen hard during the cold winter months.

During the winter the lush grassland of the North American prairie can turn into a snowy wasteland. ▼

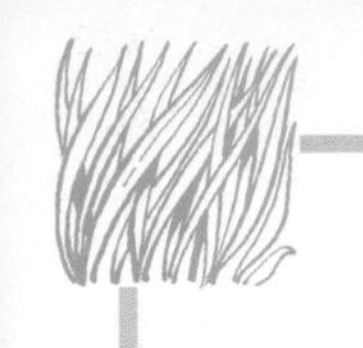

The Grassland Habitat

The most obvious feature of a grassland is a blanket of grass stretching as far as the eye can see. In tropical grasslands, the tall grasses often provide cover for predators to stalk their prey unseen. Temperate grasslands are generally made up of smaller grasses. This means that herbivores are more exposed; it is also harder for a predator to creep up on them unnoticed. In the winter, herbivores may need to trample on the snow-covered ground to reach the grass below.

▲ A typical grassland habitat

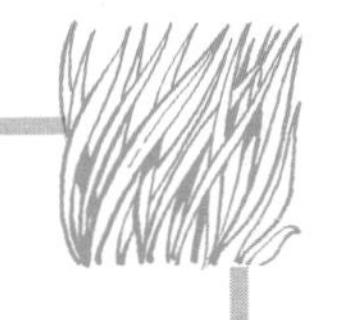

Grasses are very hardy and can survive almost all conditions except lack of sunlight. Unlike most plants, grass grows from the base of the leaf. If the grasses' leaves are nibbled down by herbivores, they can easily grow back and replace the lost leaves. Grasslands are often exposed to harsh winds. The mass of plant roots deep below the ground help prevent the soil from being blown away.

Most grasses only have tiny flowers because they do not need to attract insects to **pollinate** them. They reproduce by using the wind to scatter their seeds. ▼

▲ During the dry season in the African grassland, enormous herds of herbivores, such as wildebeests, travel great distances to find new pastures.

Grassland soil is poor quality and only a few trees are able to grow in it. Trees provide nesting places for birds and shade from the heat of the midday sun. Because many herbivores eat the seeds and seedlings, only a few trees manage to grow. Above the grasses, small insect-eating birds swoop down to catch insects. High in the sky, predatory hawks, eagles, and vultures scan the ground for food. Below ground, prairie dogs, rabbits, and mole rats live in special burrows safe from danger.

Mammals

Big cats are major predators in the grasslands. They have excellent hearing and eyesight and sharp claws and teeth, and they are very successful hunters able to catch prey larger than themselves. Big cats have sandy-colored or spotted coats to **camouflage** them while they are stalking prey.

Lion

The lion is the largest and strongest of the African cats. A group of lions is called a pride. Each group usually contains one male, several females, and their cubs. The females hunt for zebras and wildebeests to feed the pride, sometimes working together to catch their prey. The male lion always eats first. It is his job to defend the **territory** and protect the pride.

▲ The male lion has an impressive mane that makes him look even bigger and stronger than he is.

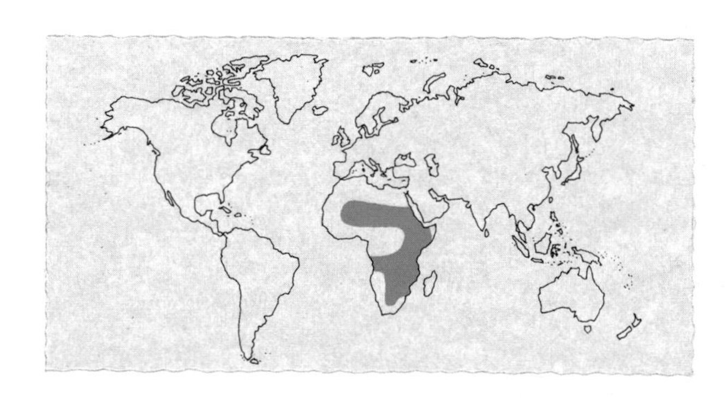

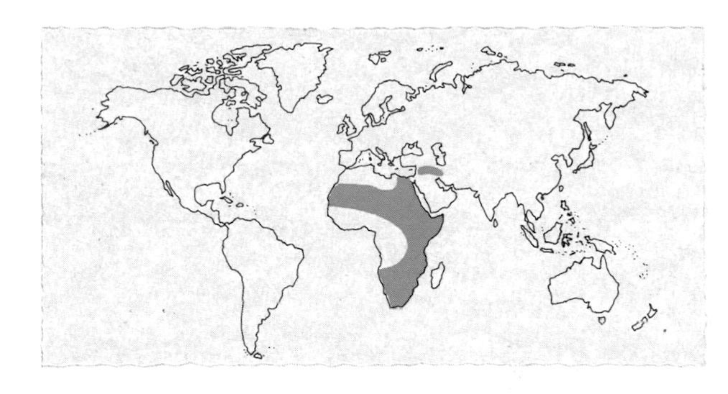

Cheetah

The cheetah is the fastest four-legged animal in the world. When chasing prey, it is capable of sprinting up to 60 mph. It can only keep up this speed for about 60 seconds before giving up the chase. The cheetah mainly catches the fast-running antelope. Male cheetahs often live together in small groups, but the females live alone unless they are rearing cubs. The female teaches the cubs to hunt, and they leave her after eighteen months.

▲ As the cheetah runs, its sharp claws help grip the ground like spiked running shoes.

Serval

Like most of the smaller cats, the long-legged serval lives alone. It has a variety of hunting techniques to catch small mammals, birds, lizards, and insects. Servals have large ears for detecting prey and often pounce on their victims with a graceful leap. The serval can leap up to 10 feet in the air to grab a passing bird. Serval cubs leave their mother and fend for themselves when they are one year old.

The serval's hind legs are longer than its front legs, which gives it extra power for leaping. ▼

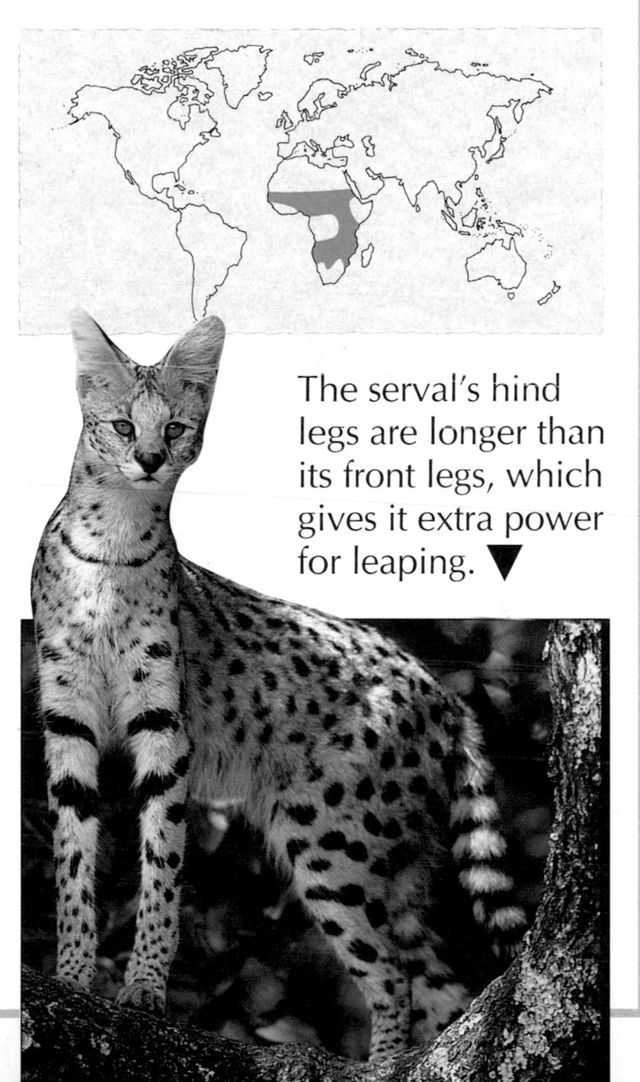

Other large and small hunters of the grasslands are African wild dogs, wolves, coyotes, hyenas, foxes, and weasels. These animals have excellent hearing, eyesight, and sense of smell, for detecting a variety of prey.

Maned wolf

The maned wolf is a shy animal that lives alone. Its favorite food is wild guinea pig, but it also eats rabbits, **rodents**, lizards, birds, and even fruit. The maned wolf wades through the grass, stopping frequently to sniff the air and listen for prey. It creeps up and pounces on its victim. Maned wolf cubs are born in a **den,** and the male brings back food to help feed them.

▲ The maned wolf has long legs, which allow it to see above the tall grass.

▲ If the hyena does not finish its food, it will bury it for another time.

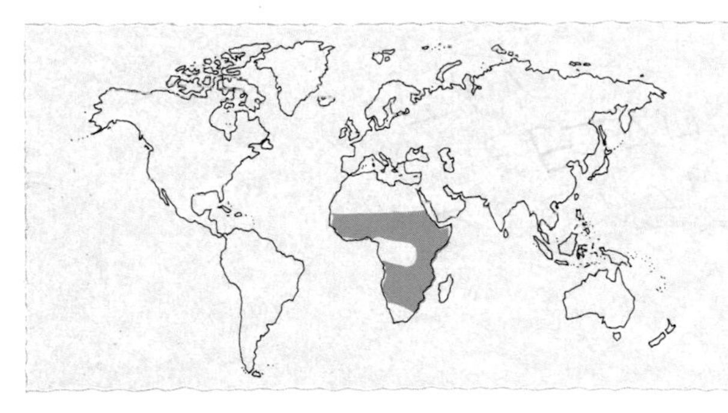

Spotted hyena

Spotted hyenas live in groups called clans, which may contain between twelve and a hundred animals. The hyena is a skilled hunter and **scavenger**, able to kill a zebra or chase a lion away from its meal. Its spine-chilling howls and cackling laughter strike fear into its prey. Hyenas work as a team and often hunt as a pack. They bring down their prey by seizing its hind legs. Unlike most **carnivores**, hyena pups can see and run soon after birth. By one-and-a-half years old, they are weaned from milk to meat.

Coyotes hunt their prey using a mixture of excellent smell, eyesight, and hearing. ▼

Coyote

The coyote is a close relative of the wolf. Single coyotes eat small or dead mammals, but may hunt as a small pack to tackle deer. Coyotes use a variety of sounds to communicate, and the coyote howl has become a symbol of the North American wilderness. There are usually six coyote cubs. They are born in a den and are brought food by their parents. They grow quickly and are fully grown at nine months old.

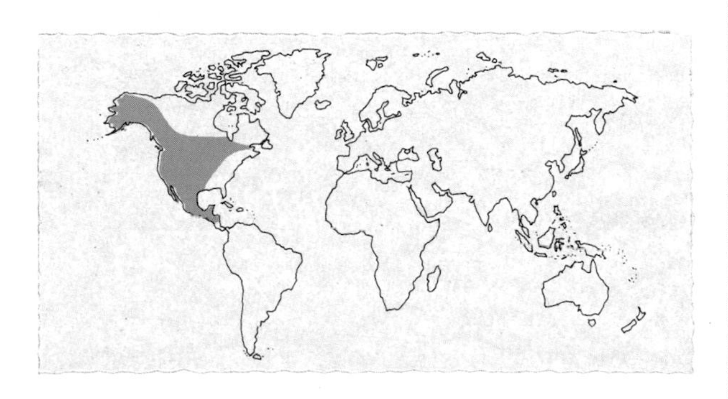

Tropical and temperate grasslands are home to a wide variety of antelope, deer, and zebra. These herbivore mammals often live in large herds for safety. They have many enemies and must keep their senses alert for an approaching predator at all times.

Grant's gazelle

The graceful Grant's gazelle is one of several types of antelopes. It feeds on grass and leaves and can survive on very little water because it extracts moisture from its food. Grant's gazelles live in separate male and female groups, sometimes mingling with herds of Thomson's gazelles. Gazelles rely on their acute senses and their ability to run fast to escape their most feared enemy—the cheetah.

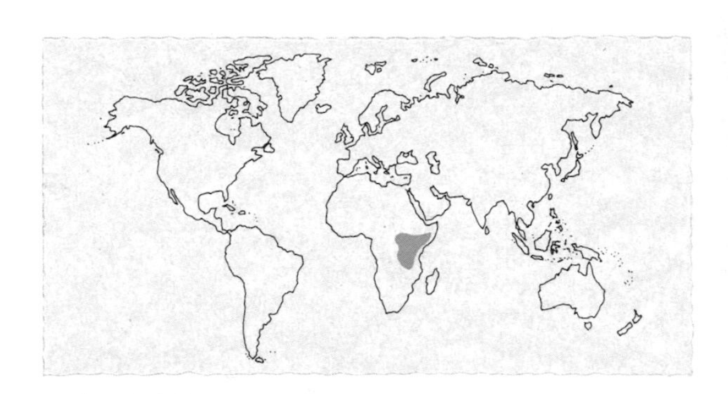

This male Grant's gazelle's large horns are a badge of strength to rival males. ▼

◀ Only the male saiga has horns.

Saiga

The saiga is an unusual-looking antelope that lives in the cold grasslands of Asia. It lives in large groups that roam the grassland in search of fresh grass, avoiding the saiga's main enemy, the wolf. The saiga's large snout contains passageways where cold, dry winter air is warmed and moistened before reaching its lungs. It is quite common for a female saiga to give birth to twins. The young saigas are suckled by the female until they are three months old.

Common zebra

The common zebra is a striped relative of the horse. It spends many hours each day feeding on grasses. Zebras live in herds and sometimes mix with herds of antelope. Zebras' stripes help confuse an attacking predator because their striped bodies blend together as they flee, making it harder for a predator to single out one animal. Zebras can also defend themselves by kicking backward with their hooves.

Every zebra has a different pattern of stripes, which is how zebras recognize one another. ▼

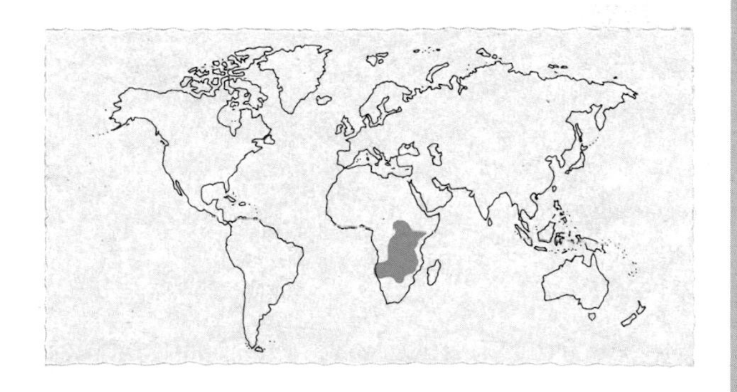

Some of the largest land mammals, including the elephant and the rhinoceros, are herbivores. Their size protects them because few predators would try and attack such huge beasts. Sadly, they have no protection against the most dangerous predator of all—humans—and many are now endangered.

African elephant

The African elephant is the largest, strongest land mammal, but it is also one of the most gentle. African elephants live in family groups of females and calves, while the males spend much of their lives alone.
The African elephant flaps its large ears to keep cool. It enjoys bathing in water, followed by a "dust-bath," throwing dirt with its trunk over its body to protect itself against insect bites. Females can give birth every four years.

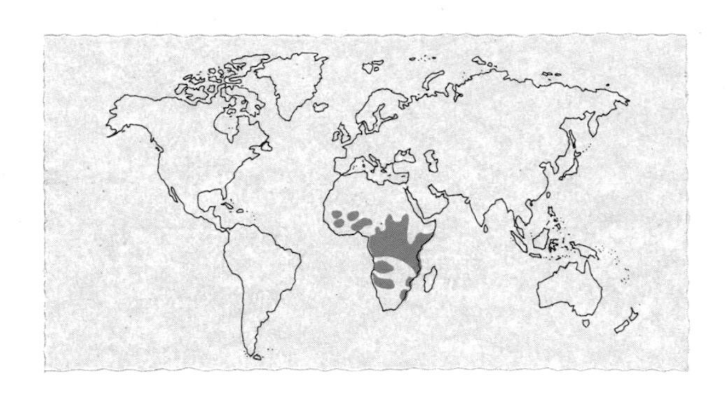

The elephant uses its trunk for breathing, smelling, picking up food, sucking up water, and greeting other elephants. ▼

▲ Neither the rhinoceros nor the elephant wanders too far away from water.

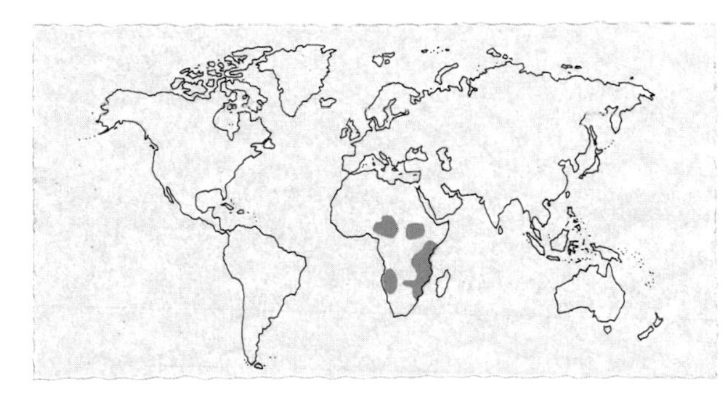

Black rhinoceros

Like the elephant, the black rhinoceros has tough skin to protect it from danger. Instead of tusks, it has two horns on its head. The black rhino mainly feeds on leaves from trees and shrubs. It has poor hearing and eyesight and uses its sense of smell to detect danger and other rhinos. The rhino feeds at night and spends the day asleep under a tree or, when it is really hot, wallowing in a mud bath.

American bison

Standing almost 7 feet high and weighing 2,000 lbs., the American bison is the largest and heaviest animal in America. It is a type of wild cow that lives in small groups or forms larger herds to protect itself against its main predator, the wolf. If bison cannot escape danger, the males form a protective circle around the females and calves. Bison almost became extinct in the 1900s due to overhunting and now only live in protected areas of the prairie.

▲ American bison grow thick winter coats to protect themselves from the cold during the winter months.

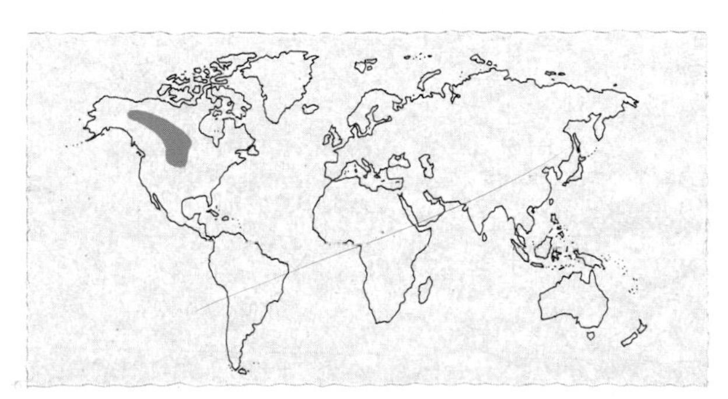

Grasslands are home to many mammals, both large and small. Some, such as the hamster and the prairie dog, live in groups in underground burrows. Others, such as the giant anteater and the **armadillo**, wander alone in search of ants and termites (*see* page 27).

Red kangaroo

Red kangaroos are the main plant-eating mammals of the Australian grasslands. They usually live in groups and may be seen bounding along on their strong back legs. Kangaroos often rest in the shade during the day and feed at night. Female kangaroos have a pouch where the joey (baby kangaroo) develops and feeds on its mother's milk. Male red kangaroos are called boomers, and females are called blue fliers.

▲ At full speed, red kangaroos can travel as fast as 40 mph. Kangaroos have thick tails that help them keep their balance.

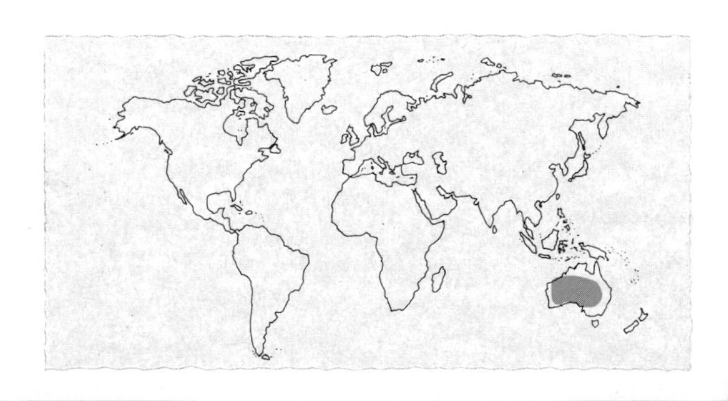

Prairie dog

Prairie dogs are related to squirrels and are named after their doglike bark. They live in large **colonies** made up of small family groups. Prairie dogs build large, underground burrow systems called "prairie towns." They live in large groups, which means there is always one dog looking out for predatory hawks and coyotes. If a prairie dog spots danger, it barks loudly and wags its tail to warn the other dogs to take cover.

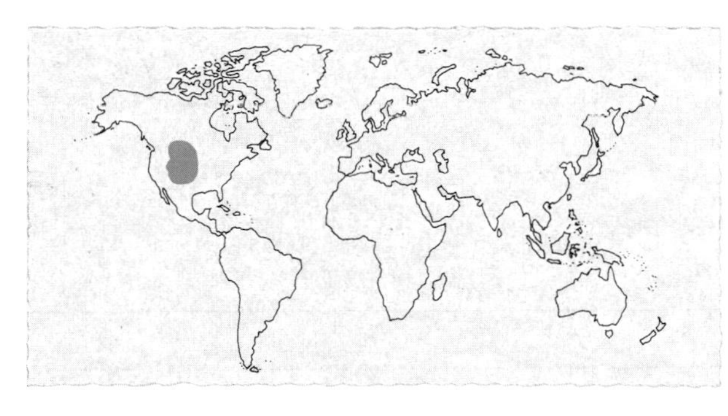

A pile of earth around each prairie dog burrow entrance serves as a lookout post and helps keep water from flooding in. ▼

▲ The male mara sits on its back legs. It is alert for any signs of danger.

Mara

The mara is a long-legged relative of the guinea pig. A male and female mara pair together for life within a large group. Their young are born in a nursery burrow, and the mother frequently returns to feed them. The female mara eats plants and grass to produce the milk to feed her young.

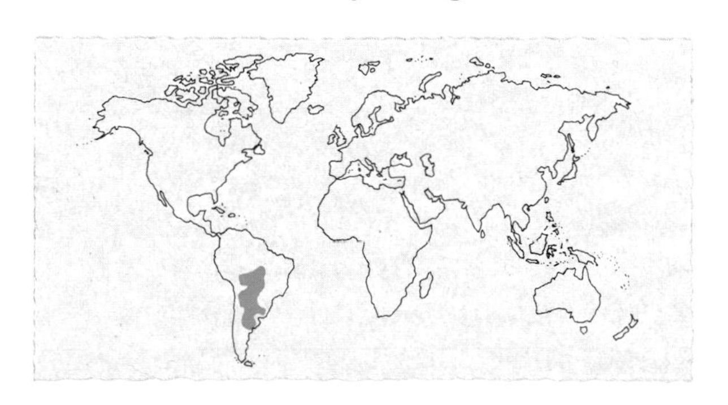

Birds

Grassland birds vary greatly in shape, color, and size. They include the ostrich, the world's largest bird, and the tiny weaver bird, which weaves its treetop nest from grasses.

Emu

The emu is a large, flightless bird that looks similar to the African ostrich. It is Australia's largest bird and lives in small flocks that roam the grassland. The emu eats plants (but not grass), fruit, seeds, insects, lizards, and rodents.

When there is plenty of food, the emu builds up fat reserves to live off when food is scarce. The male builds a bowl-shaped nest and guards the newly laid eggs. It looks after the newborn striped chicks for seven months.

▲ The emu stands 6 feet high on long, powerful legs, designed for long-distance walking.

▲ The attention of the oxpecker is welcomed by the antelope, because it feeds on irritating pests and parasites.

Oxpecker

The oxpecker is a small bird that spends much of its time riding on the backs of zebras, rhinos, giraffes, and buffalo. Oxpeckers have sharp claws for holding onto an animal as they search its body for blood-sucking parasites and flies. They warn the animal if danger is approaching by screeching loudly. Oxpeckers build their nests in holes high up in trees. They often pull out hair from large animals to use as nesting material.

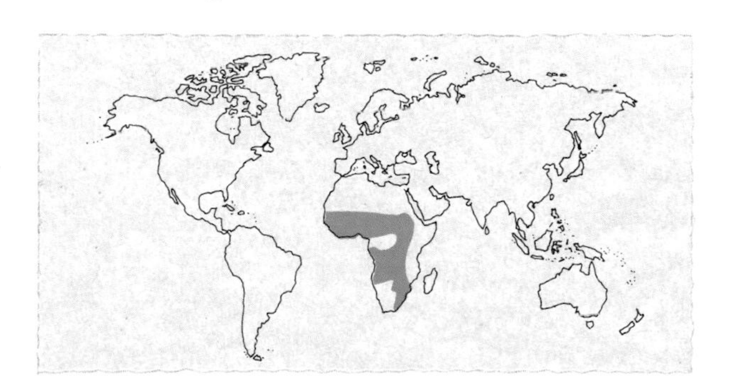

Sage Grouse

The sage grouse is named after the silvery-gray shrub called sagebrush, which the grouse eats and also uses for shelter. The sage grouse is perfectly adapted for life on the ground, but the greatest danger comes from birds of prey. In the spring, male grouse gather at special sites, where they take part in mock fights to obtain the best territory. When the female birds arrive, the males display their tail feathers. Their loud mating call can be heard a quarter of a mile away.

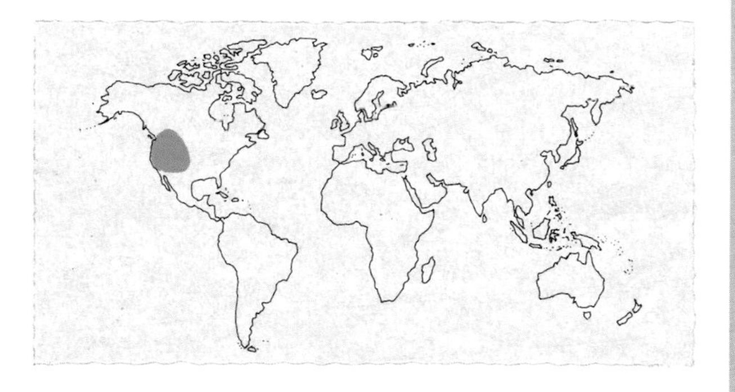

Male sage grouse erect their tail feathers and puff out their necks and throat sacs to warn off other males. ▼

Grasslands are ideal habitats for carnivorous birds, such as eagles, hawks, and vultures. These birds soar high in the sky on their large wings, searching for prey on the ground below.

▲ The vulture has good eyesight and can spot food from a distance of 1 mile.

Lappet-faced vulture

With a wingspan of 8.5 feet, the lappet-faced vulture is the largest and most aggressive of the vultures. The vulture plays an important role in the cycle of the grassland by eating the remains of other animals' meals. It feeds on **carrion**, often cleaning up the leftovers from a lion's meal. Different vultures eat different parts of an animal, but lappet-faced vultures chase away smaller vultures. They often follow the migration of grass-eating mammals, looking for a sick or newborn animal to attack.

▲ Secretary birds are named after their head crests, which look like the quill pens used by human secretaries many years ago.

Secretary bird

The secretary bird is a long-legged relative of the hawk and the eagle. Unlike other birds of prey, the secretary bird walks along the ground catching insects, small mammals, and birds with its beak. It kills larger animals, such as snakes, by stamping on them with its large clawed feet. Long legs help the secretary bird to see over the tall grass. It builds a large nest high on a flat-topped tree or thorny bush.

Golden eagle

The golden eagle is a large eagle that soars over open grassland in search of prey. It feeds mainly on small mammals and birds, but occasionally eats carrion. Once the eagle spots its victim, it half closes its wings and swoops down from the sky at 93 mph. The eagle grabs its prey with its talons and tears it into smaller pieces with its hooked beak. Golden eagles nest on rocky ledges or treetops, where the female lays her eggs.

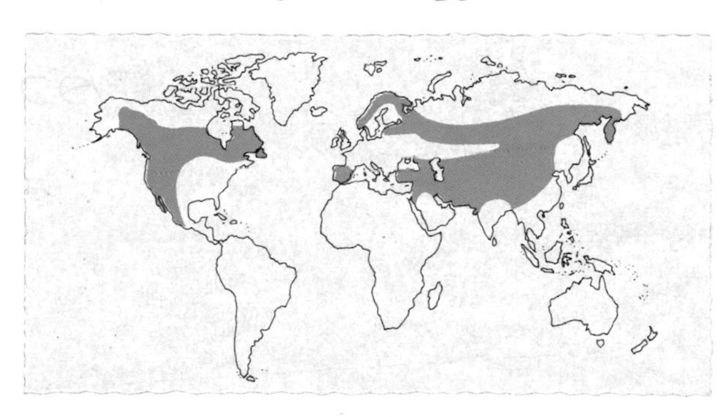

The golden eagle has a wingspan of 6.5 ft and soars easily in the air. ▼

Reptiles

Lizards, snakes, and tortoises are all **reptiles** living in grasslands. The tall grasses make it easy for snakes to stalk mammals and birds. In temperate grasslands, **cold-blooded** reptiles shelter in burrows to escape the cold winter weather.

Leopard tortoise

The leopard tortoise slowly ambles across the ground, carrying its mobile home on its back. In times of danger, the tortoise pulls its head and legs inside its tough shell and waits until the danger has passed. When a male leopard tortoise competes for a female, it pushes and shoves the rival tortoise, hoping to tip it on its back. Before the female tortoise lays its eggs, it **urinates** to soften the soil. It then digs a nest hole with its back legs and lays up to thirty eggs in the hole.

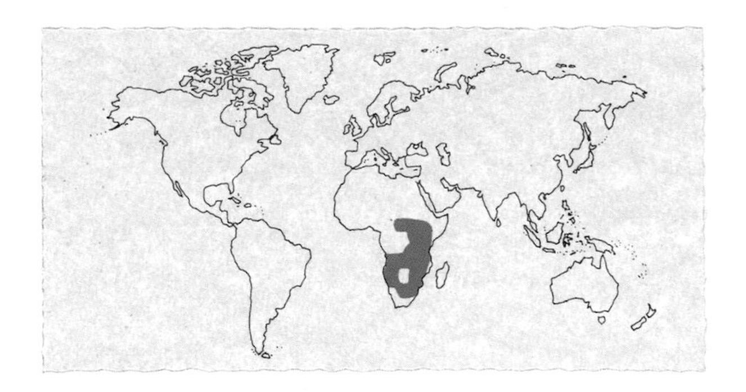

The leopard tortoise is named after its boldly patterned shell. ▼

Gopher snake

The nonpoisonous gopher snake lives in a wide variety of habitats, including grasslands. The gopher snake is usually active during the day and eats mainly rodents, but also catches rabbits, birds, and lizards. The gopher snake kills its prey by coiling its body around the animal and squeezing until its victim suffocates. In the spring, the female gopher snake digs a burrow and lays twenty-four eggs, which hatch after about eleven weeks.

When threatened, the gopher snake hisses and vibrates its tail. ▼

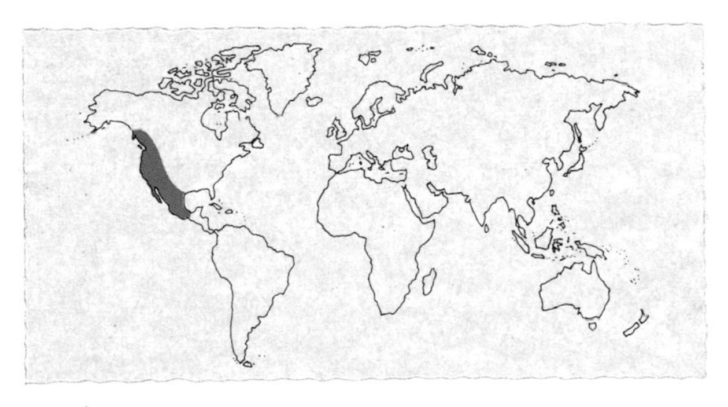

▲ An adult common tegu can be as long as 4.5 feet.

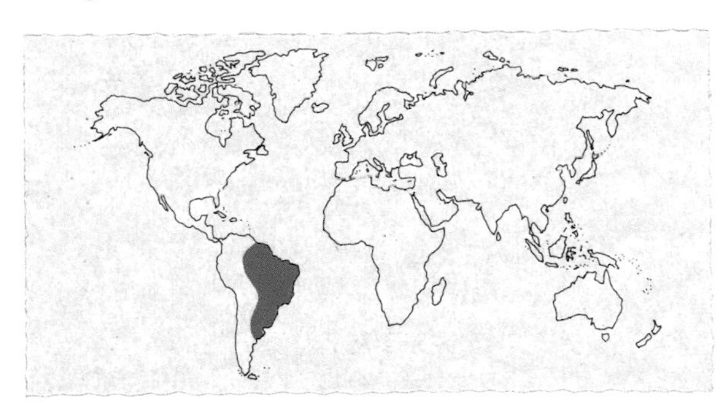

Common tegu

The common tegu is a large, ground-living lizard preferring areas with dense undergrowth. It hunts for food on warm days and spends cooler days in its burrow. Tegus eat a variety of prey, including small mammals, birds, amphibians, and insects. They also eat juicy fruits and leaves. When threatened, a tegu can defend itself by using its long, hard tail as a whip. However, they prefer to escape rather than to fight.

Invertebrates

Tall grasses form a miniature jungle for tiny **invertebrates** to live in. Some, such as the grasshopper and locust, actually eat parts of the grasses. For a web-spinning spider, the grass stems are ideal places to attach its web.

Locust

Locusts look like large grasshoppers. Their back legs are long and powerful for hopping. Locusts hatch from eggs laid in the ground and are tiny versions of the adults, but without wings. Locusts are disliked by farmers because of the terrible damage they can cause to crops. They can eat most vegetation with their powerful jaws and feed almost constantly after hatching.

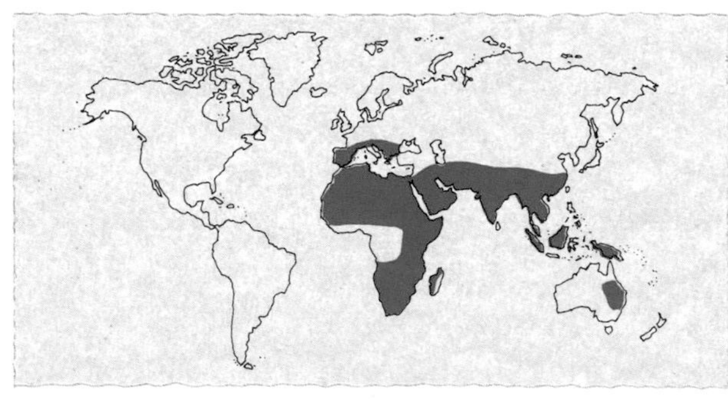

◀ A locust eats the equivalent of its own body weight in food each day.

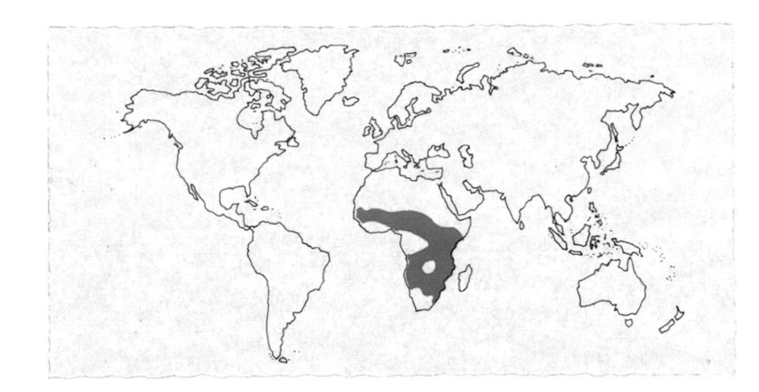

Termites

Termites are tiny, antlike insects that live in colonies containing thousands of insects. Like ants, some termites build underground nests, while others build vast clay castles made from earth and saliva. A termite nest has many chambers and air shafts that keep the nest temperature constant. Each nest contains a queen, a king, workers, and soldiers with large jaws for guarding the nest. However, termites have no defense against large predators, such as the anteater and the **aardvark**.

▲ A pair of dung beetles rolling a ball of dung

Dung beetle

If it wasn't for insects such as the dung beetle, the grassland would be covered in animal **dung**. These beetles eat the droppings of larger animals and turn them back into **nutrients** that then feed the plants. The female lays her eggs in a dung ball, which is kept soft and safe underground. The dung is food for the beetle **larvae** when they hatch inside the ball.

▲ The large, egg-laying queen termite rules the smaller worker and soldier termites from her royal chamber.

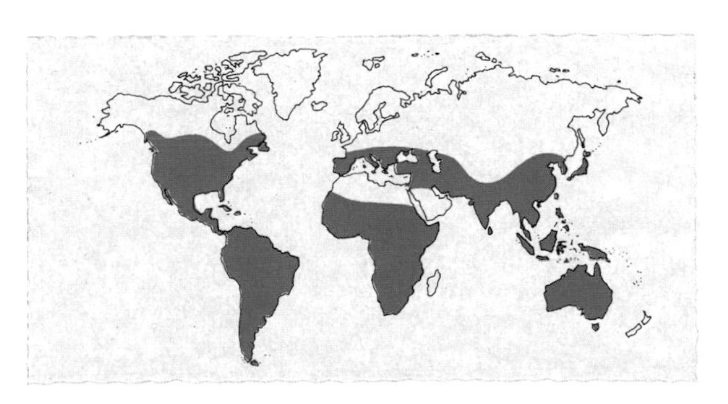

Within the tall grasses, a battle for survival goes on constantly. Butterflies spend their short lives looking for a mate, bees collect nectar, and spiders lie in wait in their underground burrows to catch prey.

▲ The red eye spots on the swallowtail's wings help distract predators from its vulnerable body.

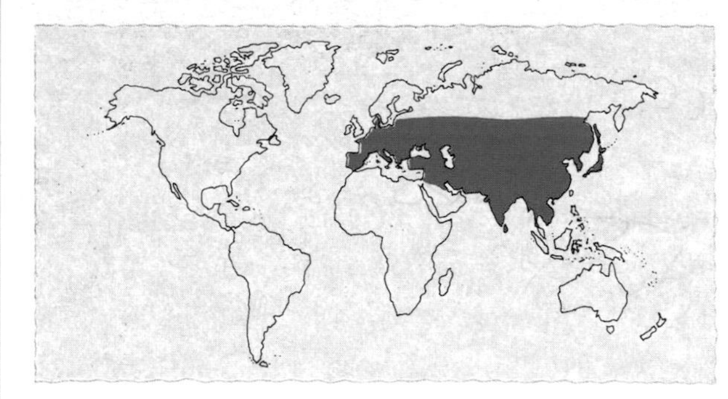

Swallowtail butterfly

The swallowtail butterfly is just one of many butterflies that live in the grasslands. The swallowtail lays its eggs on plants. After ten days, the caterpillars hatch and feed on the plants. Newly hatched caterpillars are camouflaged to look like bird droppings as a defense against predators, but they grow to look more like caterpillars. However, they can defend themselves against birds, spiders, and small mammals by inflating a pair of orange horns that give off a foul smell.

▲ The hexagonal cells of this African honey bees' nest contain developing bee larvae and honey and pollen that are used as food.

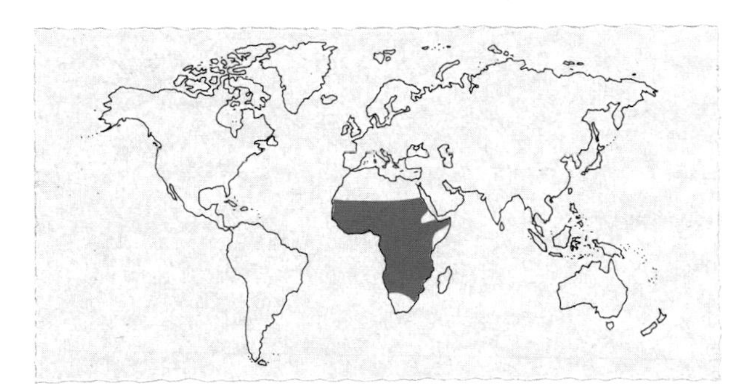

African honey bee

Wild honey bees usually nest in a tree by either hanging from a branch or nesting inside a hollow. The nest, which may contain 80,000 bees, is ruled by a single queen bee. Most of the bees are workers, who maintain the nest, collect food, and feed the young and the queen. The nest is made from hexagonal cells made of wax. The wax comes from special glands in the bees' bodies. It is chewed and then used to make the hexagonal cells.

Trapdoor spider

The trapdoor spider lives in an underground burrow, where it hides from predators and lies in wait to catch prey. The spider digs its burrow using special spines on its fangs and makes a hinged trapdoor from spun silk and soil. In grasslands, trapdoor spiders often build their burrows in the shade of a tree and catch insects attracted to the cool shade.

The trapdoor spider rushes out at its prey from beneath the trapdoor and bites it with its poisonous fangs. ▼

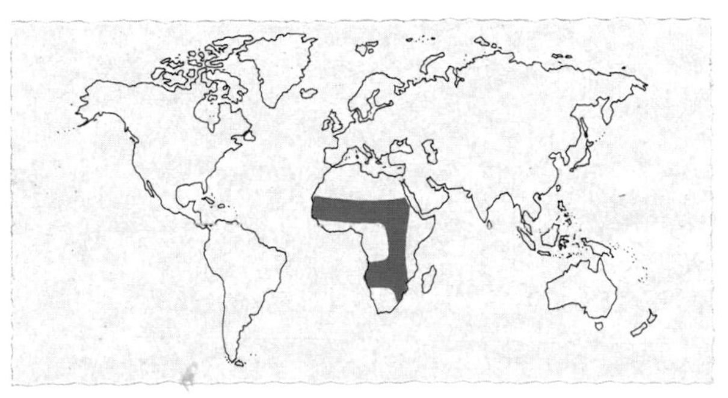

Glossary

Aardvark An African mammal with long ears and a snout that eats termites.

Armadillo A South American mammal whose body is protected by armor-plated skin.

Camouflage The way that animals escape the notice of predators, usually by matching their appearance to their surroundings.

Carnivores Animals that eat only other animals.

Carrion The body of a dead animal.

Cold-blooded Animals that are not able to make heat to warm their bodies. Their body temperature is similar to that of their surroundings.

Colonies Groups of the same kind of animal that live and work closely together. Ants live in colonies.

Den An animal's home, which may be a small cave or a hole dug in the ground.

Dung Manure produced by a large mammal, such as a cow.

Herbivores Animals that eat only plants.

Invertebrates Animals without a backbone.

Larvae The stage of development that most insects become between hatching from an egg and growing into an adult insect. A caterpillar is a larvae.

Mammals Animals that are warm-blooded. Mammals give birth to live young.

Nutrients Substances that are taken in by plants and animals to help them grow.

Pollinate To fertilize with pollen. Seeds are made from this process.

Prairie A treeless, grassy plain.

Predators Animals that hunt and kill other animals for food.

Prey An animal that is hunted or captured by another for food.

Reptiles Cold-blooded animals. Reptiles' young hatch from eggs.

Rodents Small mammals that have teeth for gnawing. A mouse is a rodent.

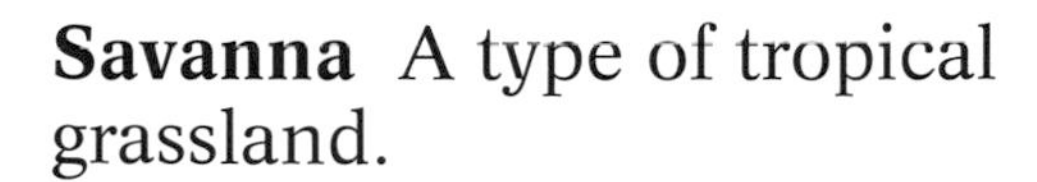

Savanna A type of tropical grassland.

Scavenger An animal that eats dead animals or other animals' leftovers.

Seasonal Something that happens during a certain season of the year.

Temperate Countries with a warm climate.

Territory An area of land where an animal or group of animals live, which they defend against other animals.

Tropical The climate of regions lying either side of the equator.

For Further Reading

Amsel, Sheri. Grasslands. Habitats of the World. Austin, TX: Raintree Steck-Vaughn, 1992.

Collinson, Alan. Grasslands. Ecology Watch. Morristown, NJ: Silver Burdett Press, 1992.

Greene, Laura O. Wildlife Poaching. New York: Franklin Watts, 1994.

Hedren, Tippi and Taylor, Theodore. The Cats of Shambala. Acton, CA: Tiger Island Press, 1992.

Sayre, April P. Grassland. Exploring Earth's Biomes. New York: 21st Century Books, 1994.

Notes About Habitats

The world is divided into various habitat types, including deserts, grasslands, rain forests, temperate forests, mountains, and oceans. The distribution of these habitats is partly determined by the topography of the land and partly by the climate. Together, these two factors help shape the face of the planet. A way of classifying habitats is by the amount of rainfall they receive.

In some parts of the world, these different habitats have distinct borders, for example, when a forest meets the sea. However, it is more common for habitats to merge slowly into one another, such as a desert merging into a grassland. Consequently, some animals may be found in more than one habitat: Caracals may be found in deserts and grasslands, and birds of prey may soar over various habitats searching for food.

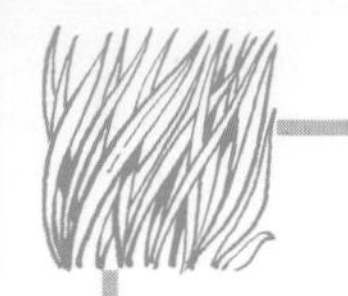

Index

Page numbers of illustrations are in bold.

birds 9, 11, 12, 20–23, 24, 25, 28
  eagles 22, 23, **23**
  emus 20, **20**
  hawks 19, 22
  insect-eating 9
  ostriches 20
  oxpeckers 21, **21**
  sage grouse 21, **21**
  secretary birds 23, **23**
  vultures 22, **22**
  weaver 20
burrows 9, 19, 28

camouflage 10, 30
carrion 22, 23, 30
climate 7

endangered species 5, 16

hunting 5, 12

insects 4, 9, 11, 20, 29
  bees 28
    honey 29
  butterflies 28
    swallowtail 28
  dung beetles 27, **27**
  grasshoppers 4, 26
  locusts 26, **26**
  spiders 4, 26, 28
    trapdoor 29, **29**
  termites 27, **27**
invertebrates 26–29, 30

larvae 27, 30

mammals 10–19, 30
  aardvarks 27, 30
  anteaters 18, 27
  antelopes **4**, 14, **21**
  armadillos 18, 30
  bison, American 17, **17**
  buffalo 21
  cheetahs 11, **11**, 14
  coyotes 12, 13, **13**, 19
  deer 13, 14
  elephants 16, **16**, 17
    African 5, **5**, 16
  foxes 12
  gazelles
    Grant's 14, **14**
    Thomson's 14
  guinea pigs, wild 12
  hamsters 18
  hyenas 12, **13**
    spotted 13
  lions 10, **10**, 13, 22
  maras 19, **19**
  mole rats 9
  prairie dogs 9, 18, 19, **19**
  rabbits 9, 12, 25
  rhinoceros 16, **17**, 21
  saigas 15, **15**
  servals 11, **11**
  weasels 12
  wildebeests **9**, 10
  wolves 12, 13, 17
    maned 12, **12**
  zebras 10, 13, 14, 15, **15**, 21

pampas 7
parasites 21, **21**
prairies 7, **7**, 17, 30
predators 4, 10, 14, 16, 17, 27, 30
prey 4, 12, 29, 30

reptiles 12, 20, 30
  lizards 11, 12, 20, 25
  snakes 23, 25
    gopher 25, **25**
  tortoises 24
    leopard 24, **24**

savanna 7, 31
scavengers 13, 31